The Drug Discovery and Development Cycle

The Drug Discovery and Development Cycle

A concise overview of the key steps from concept to launch

Kabir Hussain

The Drug Discovery and Development Cycle

First Edition

Kabir Hussain

ISBN: 9798666872178

Contents

Introduction

Once a potential drug is discovered, the road to bringing it to market is long and costly. Above all it requires strong collaboration between scientists from a variety of disciplines as well as regulatory and commercial expertise to successfully bring the drug to market. Starting a career in the pharmaceutical industry can feel a little overwhelming at the beginning, whether you have freshly graduated and chosen to embark on this career path or made a switch from academia, or even another industry entirely. When I was starting my career in the industry after graduating from university, I found that there were very few resources from which I could gain a more general overview of how the drug discovery and development process works and what it is like to work in the industry. There were comprehensive textbooks that described the intricate details of every aspect of the drug discovery and development process, but as someone who was new to the industry it was difficult to absorb such detailed information without having prior experience. Therefore, I decided to write this book to provide a concise overview of the drug discovery and development process that is hopefully a little easier to digest. The book consists of brief chapters describing each step of the drug discovery and development process (Part 1) and further chapters that expand a little on other aspects of the process and of the pharmaceutical industry in general (Part 2). I draw on 13 years of experience working in this field, in which I have had the opportunity to work for a biologics research and development company (MedImmune, formerly a subsidiary of AstraZeneca), two contract research organisations (Covance Inc. and Eurofins Pharma Bioanalytical Services) and a biotechnology company (Immunocore Limited) where I currently work as a senior scientist in the Bioanalytical department. I hope that by the time you have finished reading this book, you will have a general understanding of the drug discovery and development process and that it will help you to see where you could fit if you are considering a career in the industry.

Note: The pharmaceutical industry's main focus is on two types of drugs: small molecules and large molecule biologics. While both classes of drugs follow the same overall drug discovery and development process, characteristics that are unique to each of the two types of drugs mean that there are differences in the production and testing of each. This book aims to use both as examples to explain the drug discovery and development process. The word *drug* is used interchangeably for small molecule drugs and biologics in this book, and there is a chapter that discusses the difference between the two types of drugs (chapter 10). Further, while small molecule drugs are the focus of pharmaceutical companies and large molecule biologic drugs of biotechnology companies, the term *pharmaceutical industry* in this book is intended to cover both.

PART 1

Chapter 1:
Overview of the drug discovery and development cycle

Drug research and development is at the core of the pharmaceutical industry. The goal is to discover and develop new medicines that can help treat or even cure diseases in patients. The road to achieving this goal is long, expensive and fraught with many risks and challenges. The entire process from discovery to launch can take upwards of 10 years and cost in excess of $1 billion. Despite this, pharmaceutical and biotechnology companies take on this risk because the benefits of succeeding can be extremely rewarding, both in terms of being able to treat patients, thereby saving lives and/or improving quality of life, and from a financial perspective. There are multiple stages to the drug discovery and development process, and collectively they require a diversity of theoretical and practical scientific knowledge as well as regulatory and commercial expertise in order to overcome and reach the end goal. As such there is an immense collaborative effort in driving potential drug molecules through the process.

Figure 1 shows the keys steps in the drug development process.

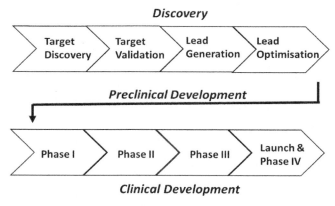

Figure 1: Key steps in the drug discovery and development cycle

The drug discovery and development process is often pictured as a linear process (as shown in figure 1), but the reality is that there is often an overlap of functions and re-evaluations at certain stages. This is because there is a constant drive to improve the drug as more and more is understood as it proceeds along the course. Similarly, not every drug molecule is developed and tested in the same way because there are so many different types of drugs that have unique attributes and as a result need to be assessed in a variety of ways.

The subsequent chapters in part 1 will focus on the drug discovery and development process by describing each stage of a typical (but not universal) route from discovery to market. The early discovery phase, in particular, will be explored in the context of a small molecule drug in order to maintain a singular focus on the process itself. The differences between small molecule drugs and large molecule biologic drugs are explored a little further in chapter 10.

Chapter 2:
Discovery

Target Identification

Target identification is the first step in the drug discovery process. Once a particular disease is chosen for potential treatment, the research team will look for biological targets that are associated with the disease. This is often a protein that is found within a micro-organism or a protein within the human body that plays a key role in the disease. The target protein may be involved in causing the disease, in which case the therapeutic will be designed to block its action. Conversely, the target protein may play a role in alleviating the effects of the disease, in which case the therapeutic may be designed to augment its action. Understanding the molecular mechanisms of the disease is critical; for example, studying the key signalling

16

pathways associated with the disease will help to identify the protein or the receptor that the intended drug needs to target.

Common drug targets include:

- Enzymes
- Cell surface receptors
- Intra-cellular receptors
- Ligand-gated ion channels
- Voltage-gated ion channels
- Gene transcription factors

Target validation

Once a biological target for the drug is identified, it is necessary to demonstrate that this target does, in fact, play a key role in the disease. This process is known as target validation, and it is vital in helping to build confidence that the appropriate target has been selected. Thus, investing time and effort in this step will enhance the probability of success further down the line.

There are a number of ways in which a drug target can be validated:

Literature – The wealth of existing knowledge through literature searches and case studies can be utilised. It is easier to benefit from this method if the drug target is one that has been widely studied; for example, it may be a molecule that is a common target of drugs that are already on the market. It is more difficult to gain an understanding of the role in a disease mechanism when the target is a novel one.

In-silico – The use of computer models to study the interactions of drug targets and their ligands is a relatively cost-effective and quick method that can be used in both target identification and validation. The limitation of computer models is that while they can provide detailed information on target-ligand interactions, they cannot on their own provide enough information on these interactions in the context of living biological systems.

Gene editing – This is a common method used to see what happens when the gene encoding the drug target is altered. This can be done by either knocking down the gene, which essentially reduces the amount of protein that is expressed (achieved by interfering with the process of mRNA being translated into protein) or by knocking out the gene entirely so that the protein cannot be produced at all. The effect of this can then be studied on live cells or in live animal models. Conversely, the gene can be manipulated to overexpress the target in order to study the effects of disease initiation and/or progression as a result of this.

Proteomics – This involves modulating the activity of the target protein (as opposed to interfering with the expression of the protein described in the previous point). One example of this is to create antibodies directed against the target protein and use them to bind to the functional site of the protein, thereby blocking its action. The effect on the disease progression in a cell can then be studied. One major advantage of this approach over a genomics-based approach is that it looks at the particular protein of interest, whereas gene editing can result in isoforms of the protein that may not necessarily be identical to its naturally occurring form.

Biomarkers – Biomarkers are naturally occurring biological substances or characteristics that can be measured to identify/evaluate a particular pathological or a normal physiological process in an organism. Biomarkers are commonly used in both the early drug discovery phase as well as later development phases (this will be discussed in later chapters), but in the context of target validation, biomarkers can be evaluated to measure their reaction to changes in their direct or indirect exposure to the drug target.

These examples represent a handful of methods by which a target can be validated, but this is by no means a comprehensive list. The range of methods available for validation of a target largely depends on the type of drug being considered in the first place. Nevertheless, it is important to validate the drug target as much as possible in order to improve the chances of success further down the process. Once sufficient data has been collected and there is enough confidence that it is a true and effective target for the disease, the next step is to identify potential drug candidates.

Hit identification

Once a target for a potential therapeutic has been identified and validated, the next step is to find molecules that will bind to this target. There is a range of sources for potential drug compounds, but a drug company typically has a vast collection of compounds that are stored in what is known as a 'compound library'. Such collections of compounds can number in the hundreds of thousands and so a great deal of care and attention is taken to ensure they are appropriately stored and catalogued. A typical compound library consists of many shelves containing bottles of compounds stacked on one another, and the sheer quantity of compounds often requires automation of the retrieval of samples for screening, through the use of robotic arms. Some companies maintain compound libraries for the purpose of selling access to others, making the library a valuable commodity in itself. A drug company may also source compounds from academic institutions through collaboration or even synthesise the desired compounds afresh.

In any case it is necessary to be able to screen vast libraries of compounds quickly and efficiently in order to identify potential candidates that will meet the initial criterion of binding to the intended target. This is achieved through a well-established method known as high throughput screening (HTS) and uses a combination of robotics, liquid handling instruments and data processing software to test compounds for various properties that may make them suitable candidates for the drug target. Depending on the design of the particular HTS assay, thousands of compounds can be screened simultaneously, and millions can be screened over a short period of time. Below is a summary of some beneficial characteristics of HTS screening:

Highly automated – Use of microplates with thousands of wells (for example, 1536, 3456 or 6144-well microplates) in conjunction with liquid handling systems means that manual handling is kept to a minimum.

Efficient – The automation aspect means that the process is quick and cost effective compared to screening such large numbers of compounds manually.

Highly robust – Well developed assays can be highly robust, thereby giving confidence in the reproducibility of the procedure, which is critical when such a large number of compounds are being tested for the same purpose.

Versatile – Can be applied to a range of different assay types (for example, biochemical, cell-based and bacterial assays).

The types of assays available for screening compounds can be divided into those that measure/detect a phenotypic or physiological response in cells, and those that measure or detect interaction of the compounds with the drug target itself, that is, biochemical assays:

Biochemical (in-vitro) assays are designed to help understand the activity of biological molecules such as enzymes. By shedding light on the impact of the compounds on enzymatic activity, these assays help to separate promising compounds from those that are unlikely to have any valuable biological activity.

Cell-based (in-vivo) assays are considered to be a more effective method for screening compounds as they demonstrate biological activity in living cells, thereby providing more quality hits since they give a better indication of potential efficacy further down the line. These assays can measure the various effects of the compounds on the cells such as proliferation, changes in morphology, signalling pathways and cytotoxicity, to name a few examples. While testing compounds in cell-based assays is not fully predictive of how they will behave in animal models, it is one of the most effective and ethical ways to screen millions of compounds for in-vivo biological activities before the use of live animals is needed for consideration.

Compounds that demonstrate the intended activity in these assays are referred to as 'hits' and the process is called 'hit identification'. An initial screening of compounds is valuable in identifying hits in the first instance, but it is necessary to further refine the number of compounds through a combination of assays using HTS (some of which are mentioned above). This helps to narrow down the number of potential drug candidates, ensuring that only the most promising ones are advanced to the next stage of the process. This is all the more crucial considering that subsequent steps in the drug discovery and development process are time consuming and costly.

It is important to note that therapeutics normally fall into two main categories: large molecule biologic drugs and small molecule drugs. The hit identification process described thus far is mostly relevant to small molecule drugs. Large molecule drugs, also known as biologics, are generated in a different manner as they are manufactured using living systems such as microorganisms or plant/animal cells. Another common class of biologics is monoclonal antibody therapeutics, which are generated using hybridoma technology. This involves injecting an animal (such as a mouse) with the antigen and then harvesting the antibody-producing B cells that are raised in response. These B cells are fused with immortal myeloma cells, resulting in a new hybrid cell line that can produce unlimited quantities of the same antibody as long as it is adequately maintained.

Thus, the process of identifying or generating 'hits' is not the same for the two classes of therapeutics. Differences between small and large molecule therapeutics are discussed in more detail in chapter 10.

Lead generation

Following the determination of potential drug candidates during hit identification, it is necessary to assess these molecules further in order to narrow down the number of potential candidates to those that not only bind to the target but have the desired biological effect. This is to ensure that the subsequent stages of the drug development process are not overburdened with too many candidate molecules, which would otherwise prove too costly and time consuming to progress. This is the purpose of the lead generation step, and it helps to improve the chances of success further down the line as it identifies the most promising candidate molecules to advance along the process. Structural properties of the molecules are assessed, and a range of considerations will determine which ones are taken forward as lead compounds. Some of these considerations are shown below:

- Indication of biological activity by demonstrating binding to target. At this stage the strength of binding (or potency) is not a concern as this can potentially be improved in the subsequent stage (lead optimisation).
- Have structural properties that make the molecule amenable to alterations by chemists in order to improve important qualities (such as the aforementioned potency).

- Properties that may affect the bioavailability (proportion that enters the blood circulation when administered to the recipient) of the molecule such as hydrophobicity or polarity.

- Presence of toxic groups, or groups that can produce toxic by-products when introduced to the body. There is a good understanding of chemical groups that can be toxic, so it is relatively easy to screen out molecules that contain such groups.

- The reversible/irreversible nature of the molecule's interaction with the target is taken into consideration. Some drug molecules are required to be reversible in their binding to the target to ensure that the intended effects of the drug are not permanent (for example, drugs that help induce sleep), while for other drugs, it is advantageous to have irreversible binding to the target (for example, some antibacterial drugs).

- Lead generation is a critical step in the process, so pharmaceutical companies often have entire teams dedicated to this process.

ADME/PK

Another valuable tool in identifying lead molecules is to conduct ADME (absorption, distribution, metabolism and excretion) and PK (pharmacokinetic) studies so that those with the best chance of succeeding further down the drug development process are identified at this early stage. Traditionally, these types of studies have been conducted in subsequent stages of drug development in large part due to the cost of conducting such studies; however, it has been recognised that a high attrition rate of drug candidates further down the line has been partly due to poor ADME/PK profiles that could otherwise have been identified earlier in the process. Thus, it is not uncommon for drug companies to invest in ADME and PK studies at the lead generation stage.

Absorption – This is the process by which a compound reaches its target tissue. Often the compound is delivered there via the bloodstream after being absorbed by the intestine. There are certain properties of the chemical that will influence the effectivity of this absorption route, such as solubility, resistance to breakdown by stomach acid and how readily it can permeate the intestinal wall. Thus, absorption plays an important role in the bioavailability of the drug (discussed further in chapter 3). If a compound does not absorb effectively by the initially chosen route of administration, then other options can be investigated, for example, intravenous (injection into the veins) or subcutaneous (injection into the tissue layer between skin and muscle) administrations.

Distribution – This is the dispersion of the drug throughout the fluids and tissues of the body. Once the compound enters circulation, it needs to then be able to reach its target site. For example, an anti-cancer drug will not be of use if it cannot find its way to the target tumour cells. There are various factors that will affect the distribution of a drug to its target site:

- Passive or active transport. Passive transport requires no energy as the compound moves in the direction of the concentration gradient. Conversely, active transport moves the compound against the concentration gradient and accordingly requires energy to do so.
- Drugs may have an affinity for plasma proteins (referred to as plasma binding proteins), which means that a proportion of the administered drug will be bound to these proteins and less will be available to the target site.

- Vascularity of the target organ will affect the amount of drug that reaches it. The more blood vessels that reach the organ, the better the chance that more drug can be delivered there.

- Distribution of drug to certain organs may be hindered by the body's natural defence mechanism. For example, a drug that targets the brain will have to overcome the blood–brain barrier.

Metabolism – This describes the breakdown of the drug into smaller compounds (metabolites) once it enters the body. This normally renders the drug inert and so it is no longer active. Drugs (in particular, small molecule drugs) are commonly metabolised in the liver, although this can also happen in other organs such as the kidneys or the intestines. Metabolism can sometimes mean that the drug is transformed rather than broken down, for example, by having other molecules added to it so that it is no longer in its parent form. This can make the drug molecule more hydrophilic, so it is easier to excrete.

Excretion – This describes the removal of the drug and its metabolites from the body. Excretion is essential to ensuring that drug and drug metabolites do not accumulate to toxic levels. There are three main routes of excretion:

Renal – excretion in the kidneys through urine.

Biliary/faecal – starts off in the liver and passes through to the gut and is finally excreted in the faeces.

Respiratory – excretion via the lungs through the action of breathing.

While these are the main routes of excretion, drugs and their metabolites can also be excreted through other bodily fluids such as saliva, tears and sweat.

Lead optimisation

In order to be considered as potential therapeutic drugs, lead molecules need to have certain properties that make them suitable for the intended purpose, for example, solubility, efficacy (the ability to produce the desired effect) and safety when administered to the patient. The lead molecules may not initially have these properties, but through lead optimisation they may be altered to attain them. It is hoped that following this process the lead molecules will specifically affect the targeted biological pathway in the disease and minimise as much as possible its interaction with other biological pathways, thereby reducing the likelihood of unwanted side effects. An organic chemist can make structural changes to the lead molecules in order to

36

introduce or enhance these valuable characteristics.

Sometimes the activities in lead optimisation overlap with those in lead generation; it depends on the approach that a given company decides to take. Activities such as ADME studies may be conducted here for the first time or revisited from lead generation following improvements to the drug in the lead optimisation process. In any case, the goal is to have at least one candidate drug that is suitable to take into the next step of the drug discovery and development process – preclinical development.

Chapter 3:
Preclinical development

The lead molecules are whittled down to as little as one candidate, which is then tested for efficacy and safety (toxicity). This stage is known as preclinical testing and normally utilises a combination of in-vitro testing and in-vivo animal models. The objective of this stage is to demonstrate as much as possible that the candidate molecule will be safe for administration in humans. While it is by no means a guarantee that the behaviour of the drug in an animal and the animal's response to it will correlate to humans, it is the only way to test a drug in a live model prior to testing in humans. As one of the main aims is to test for the likelihood of unwanted side effects of the drug, it would be unethical to perform these studies in humans. For this reason, preclinical studies are mandated by regulatory authorities, such as the

FDA (Food and Drug Administration) in the US, which use the outcome of preclinical pharmacological and toxicological studies to decide on whether or not to permit the use of the drug in clinical trials.

Preclinical testing can encompass a range of different studies and the extent to which these are conducted depends on the individual circumstances of the drug. The following are examples of assessments that may be included in preclinical testing;

PK/PD – PK/PD (pharmacokinetic/ pharmacodynamic) studies help to explain the dose-response relationship of the drug and have multiple benefits in preclinical development and later in clinical development. Pharmacokinetics is an assessment of how the drug moves through the body by absorption, distribution, metabolism and excretion. The results of a PK study can also be pivotal in helping to design a suitable dosing regimen for subsequent clinical studies.

Equally important is PD, which is an assessment of the physiological and biochemical effects of the drug in the body in relation to its concentration. Whereas in a PK study the aim is to measure the concentration of the drug, in a PD study we are measuring a biomarker, which is a biochemical or physiological change in the body that is brought about by the administration of the drug. Biomarkers are critical to helping us understand if a drug is having its intended effect or not. Biomarkers can be chemicals that are affected one way or another by the disease in question; for example, the protein β2-microglobulin is a protein that is normally present in the blood and in small amounts in the urine, but in kidney disease the concentration of β2-microglobulin increases in the urine due to damaged renal tubules that would normally reabsorb the protein. Therefore, an elevated level of β2-microglobulin in the urine is a biomarker for kidney disease. Sometimes a biomarker is not a chemical, but a physiological state; for example, changes in blood pressure can be used as a biomarker for cardiovascular disease. Figure 2 shows examples of biomarkers for a range of diseases.

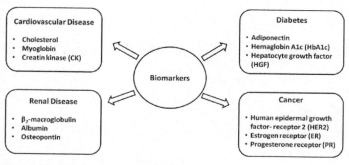

Figure 2: Examples of various disease biomarkers

A simpler way to describe PK and PD is that PK is the study of what the body does to the drug, while PD is the study of what the drug does to the body.

In a typical preclinical PK/PD study, the drug is administered to an animal species (often a rodent or a non-human primate) and then blood samples are collected at various time intervals, starting from a few minutes after dosing up to several days or even weeks (a sample is also taken prior to dosing). The samples are then analysed in the laboratory using bioanalytical assays, and the concentration of the drug (PK) or, if available, a quantifiable biomarker (PD) is measured at each timepoint. This provides a data set that can then be analysed through PK/PD modelling, which generates a set of parameters that explain the dose/concentration-response relationship of the drug. It can aid in predicting the PK/PD response in humans and is, therefore, a valuable tool in aiding the design of clinical PK/PD studies further down the line.

Acute toxicity – These studies help to assess the short-term adverse effects of the drug following single-dose administration or multiple-dose administration over a short period of time (usually around 24 hours). The study is normally carried out in two mammalian species, typically one rodent and one non-rodent species, and the effects are observed for two weeks. Various tests to measure pathological, biochemical, morphological and histological changes are performed. Acute toxicity studies help to predict the potential outcome of accidental overdosing of patients in the clinic. In some cases, one of the main purposes of an acute toxicity study is to determine the median lethal dose, or LD_{50}, that is, the dose that kills half of the animals in the dose group and has traditionally been one of the main measures of acute toxicity. However, this assessment is not as widely recommended as previously since there are ethical implications of using large numbers of animals to such ends. In addition to this, the fact that there is a lot of biological variability between animals and humans further questions the value of such an assessment.

Chronic toxicity – These studies aim to evaluate toxicity from long-term exposure and can last up to two years with repeated dosing of the drug. Typically these studies are performed in two different mammalian species (as with acute toxicity studies, this is usually done in one rodent and one non-rodent species) and often consist of three different dose levels: a high dose that produces mild toxicological effects, a low dose that is around twice the anticipated maximum clinical dose and a medium dose level in between the high and low doses. During the course of the study, the animals are carefully observed for any changes in their physiology or behaviour and also any changes in their biochemical parameters. At the end of the in-life phase, the animals are sacrificed and their tissue is harvested for histological assessment.

Reproductive toxicity and teratogenicity – These studies assess the impact of the drug on the reproductive systems in male and female animals, commonly rodents and the teratogenicity (potential to cause fetal abnormalities during pregnancy) of the drug. Males and females are dosed with the drug over a period of time and then mated, after which various assessments are made on each sex (spermatogenesis in males, and fertilisation, follicular development, implantation and fetal development in females).

Immunotoxicity – An assessment of the potential adverse effects of the drug on the immune system, such as immunosuppression, which could reduce resistance to infections or autoimmunity, resulting in the patient's immune system attacking their own body.

Carcinogenicity – An assessment of the carcinogenic potential of the drug. Such studies allow us to understand the potential of the drug to cause tumour formation in animals and to use this information to evaluate the risk in humans.

45

Any indication of carcinogenicity usually requires extensive testing in animals for at least a couple of years.

Bioavailability – A study of the proportion of and the rate at which the active pharmaceutical ingredient (API) from a drug product that reaches its intended site of action once administered. Since it is difficult to determine the concentration of a drug at its intended site of action, bioavailability studies often make this assessment by measuring the concentration of the drug in circulating blood on the basis that its concentration in the blood is equivalent to its concentration at the intended target site. Thus, bioavailability is often assessed as part of PK studies. For drugs that are administered intravenously (injected into a vein), the bioavailability is considered to be 100 per cent as the drug is directly entering the circulatory system. For non-intravenous administrations, such as subcutaneous or oral administrations, the bioavailability will not be 100 per cent as some of the drug will be metabolised and eliminated before reaching the site of action. Bioavailability can be affected by the physicochemical properties of the drug, such as particle size, as well as physiological factors relating to the patient, for example, age, sex, body weight and disease state.

Bioequivalence – A study comparing the biological similarity between two drugs in relation to their bioavailability. Sometimes a change in the drug product (for example, its formulation), its production process or even the route of delivery is required to improve its effectiveness. Bioequivalence studies allow us to determine if the change yields the same efficacy and safety properties of the drug. For companies that produce generic versions of a drug, bioequivalence studies help to demonstrate the generic's comparability to the innovator (the original drug being copied).

Chapter 4:
Clinical Development

Investigational new drug (IND)

Once the preclinical studies are completed, the next stage is to proceed to testing in humans (clinical testing). However, before this can be done, data from the preclinical studies along with manufacturing information must be collated to form an investigational new drug (IND) application. In the US, an IND is submitted to the FDA so that the agency can review and approve the use of the drug in clinical trials and allow it to be transported across state lines. As such, IND approval is considered to be a major milestone in the drug development process. The FDA states that the IND should cover the following three areas:

Animal Pharmacology and Toxicology Studies – Preclinical data to permit an assessment as to whether the product is reasonably safe for initial testing in humans. Also included are any previous experience with the drug in humans (often foreign use).

Manufacturing Information – Information pertaining to the composition, manufacturer, stability, and controls used for manufacturing the drug substance and the drug product. This information is assessed to ensure that the company can adequately produce and supply consistent batches of the drug.

Clinical Protocols and Investigator Information – Detailed protocols for proposed clinical studies to assess whether the initial-phase trials will expose subjects to unnecessary risks. Also, information on the qualifications of clinical investigators--professionals (generally physicians) who oversee the administration of the experimental compound--to assess whether they are qualified to fulfil their clinical trial

49

duties. Finally, commitments to obtain informed consent from the research subjects, to obtain review of the study by an institutional review board (IRB), and to adhere to the investigational new drug regulations.[1]

[1]https://www.fda.gov/drugs/types-applications/investigational-new-drug-ind-application

In the European Union, there is a similar process of collating information obtained thus far pertaining to the drug into a document known as the investigational medicinal product dossier (IMPD). The IMPD is submitted to the EU regulatory authorities, which will then decide on whether or not to approve the use of the drug in clinical trials, based on the information provided.

Collation of both the IND and IMPD requires extensive collaboration between scientists that have been working at various stages of the drug discovery and development process thus far. Individuals with regulatory expertise also play a crucial role in ensuring a successful IND/IMPD submission. Study reports are written, reviewed and compiled into the master document and there is great anticipation of a favourable outcome once the IND or IMPD is submitted to the relevant regulatory authority. The regulatory authority may request further work to be performed if the initial submission is not fully satisfactory, in which case the company will go back and perform the additional work and then resubmit in the hope that approval will be granted.

On successful IND or IMPD submission and approval, the next major phase in the drug development process is the clinical trials phase.

Clinical trials

The drug is first introduced to humans in the clinical trials and although there is still a long way to go and a lot of challenges to overcome, this is a major milestone along the process. The clinical trial studies are carefully designed using all the knowledge obtained up to this stage. The pharmaceutical/biotechnology company is referred to as the 'sponsor', which is the formal terminology assigned to an organisation or individual that is responsible for the initiation, supervision and financing of a clinical trial.

Embarking on a clinical trial is a considerable undertaking and the sponsor will often engage with contract research organisations (CRO) to help conduct and manage the study (CROs are often engaged with during discovery and preclinical development stages as well). CROs can be responsible for designated portions of the study, but the overall responsibility lies with the sponsor.

Careful consideration is given to the dose levels, selection of clinical trial sites and CRO partners to collaborate with, among other things. Once all the necessary preparations are complete and the volunteers/patients are recruited, the clinical trial study can commence.

Clinical trials are commonly described as being conducted over three phases (phase I, II and III); however, these are sometimes bracketed on either end by phase 0, which precedes phase I, and phase IV, which follows phase III. The next chapter will discuss the key elements of each phase individually.

Phase 0

Phase 0 studies are conducted to gain an initial understanding of how the drug behaves in the human body and how the body, in turn, reacts to the drug. These studies are exploratory by nature and differ from the other phases of clinical trials in that they are not a regulatory requirement and are hence voluntary. Phase 0 studies usually involve administering very small doses in a limited number of healthy volunteers (around 10 to 15), or in some cases patients with the disease (for example, when the drug is targeting an oncological disease). They may help answer questions like:

- Does the drug reach its intended target site?
- Is the drug adequately absorbed?
- Are there any adverse side effects from sub-therapeutic doses?
- Do cancer cells respond to the drug (oncology therapeutics)?
- What are the pharmacokinetic and pharmacodynamic properties of the drug in humans?

Despite the rigorous testing that is conducted at the preclinical stage to answer these questions, animal models can never predict with absolute certainty how a drug will behave in humans. As such there is still a very possible risk that the drug may fail to demonstrate similar safety and efficacy profiles in humans as those seen in animals at the preclinical stage. These phase 0 studies, which are very limited in scope, help to mitigate the risk of the drug failing in the main clinical trials, which is normally a time-consuming and costly endeavour in itself.

Phase I

Where phase 0 studies are not conducted, phase I studies mark the first time that the drug is tested in humans. The aim of phase I trials is to determine the highest dose of a drug that can be safely administered to humans without causing serious side effects. As mentioned previously, while the data from preclinical trials is valuable in understanding the toxicological effects in animals, the results are not guaranteed to correlate to humans.

A group of healthy volunteers are initially given a low dose of the drug to see if there are any adverse effects. The dose is then increased each time in subsequent groups of volunteers until the clinicians find a dose that is effective without causing any serious side effects. This approach is referred to as dose escalation, and there are two types of studies that can be conducted to this end:

- Single ascending dose (SAD) studies – each volunteer is given a single dose of the drug in a given group, and the dose level increases with each successive group.
- Multiple ascending dose (MAD) studies – each volunteer is given multiple doses of the drug over a predetermined period of

time, and the dose level increases with each group.

The PK and PD characteristics (previously discussed in chapter 3) of the drug are evaluated in these studies and help to design the dosing range and schedule for subsequent clinical phases. Another aspect that needs to be considered is immunogenicity, which is an assessment of anti-drug antibodies that may be generated by the body in response to treatment. Immunogenicity is discussed in more detail in chapter 11.

If volunteers from phase I show tolerance to the drug without any severe side effects, then following the appropriate review and approval of the results, the clinical trials can proceed to phase II.

Phase II

The drug is now ready to proceed to phase II of the clinical trials, in which its effectiveness in the target patient population is assessed. Whereas in phase I the volunteers receiving the drug are typically healthy individuals, phase II focuses on patient volunteers who are suffering from the target disease indication. Consequently, recruiting volunteers for the trial is one of the early challenges of a phase II study. This is especially the case when the disease is less common as there is a smaller pool of patients to recruit from. In addition, some trials may have strict eligibility criteria for patient recruitment, depending on the characteristics of the drug and information obtained from phase I, which may indicate that certain demographics of patients may be at increased risk. Phase II is also generally larger than phase I, and the number of volunteers needed can be in the hundreds.

Phase II Trials are sometimes split into two sub-phases: phase IIa and phase IIb. In this case phase IIa is designed to find a dose level that has the desired effect of treating the disease. This dose level is then used in phase IIb to confirm the efficacy, which is the principal aim of phase II studies. As with phase I, the safety of any dose administered in phase II is also closely monitored.

Phase II trials are sometimes conducted as randomised placebo-controlled studies, in which some patients receive the drug and others receive a placebo control, which is a treatment that looks like the drug but does not contain any active pharmaceutical ingredient. The purpose of this is to demonstrate that the drug has an effect compared to a control that is administered in the same way. These studies can either be single-blinded studies, in which the patients do not know if they are receiving the drug or placebo, or double-blinded studies, in which neither the patients nor the investigators know which patient is receiving which treatment. The latter is considered more robust as this reduces the potential for bias when interpreting the results.

If the phase II trials demonstrate sufficient efficacy of the drug at a safe dose level, then the clinical trials proceed to phase III, the final phase before product launch.

Phase III

Now that the drug has been shown to be effective in the patient population, it needs to demonstrate this efficacy in a larger patient population. Phase III trials can last several years and involve several hundred to several thousand patients in multiple countries. This means that multiple test sites are required, which adds further cost, time and complexity to the set-up of these trials. The logistics of organising and conducting the trial are very challenging and require many individuals and organisations to collaborate effectively, from the sponsor's scientists to the CROs, the doctors and clinicians, the regulatory authorities and those in charge of the drug supply chains to name some stakeholders.

While the main purpose of a phase III trial is to demonstrate efficacy of the drug in a larger patient population over a longer period of time, there is further valuable information to be gained from these trials:

- Continued monitoring of the safety profile of the drug and monitoring for any side effects resulting from prolonged treatment with the drug.
- Comparison of results to current treatments to see if the drug is likely to provide improved benefit over competitor drugs.

Phase III trials are often conducted as placebo-controlled double-blinded studies. If phase III trials demonstrate favourable results, then the sponsor can begin the process of seeking regulatory approval to launch the drug on to the market. Sometimes the phase III trials can be ongoing during the regulatory approval process to allow patients continued access to a potentially life-saving drug that could otherwise be interrupted until the regulatory submission and approval process concludes. Other reasons for continuing the trials include obtaining further safety information and to demonstrate efficacy in other disease indications in addition to the target disease (this is known as 'label expansion').

Chapter 5:
Regulatory submission, approval and launch

The final step prior to launching a new drug is to prepare a regulatory submission for the relevant authority, detailing the full development story of the molecule. This submission details the preclinical and clinical study findings as well as the characteristics of the drug itself along with the manufacturing and packaging process. Depending on the type of drug, the regulatory submission in the US will either be a new drug application (NDA) if it is a small molecule drug, or a biological licencing application (BLA) if it is a large molecule biologic drug. The IND (discussed in chapter 4) forms part of the NDA/BLA, which is a live document that can be updated while the clinical trials are ongoing. The FDA will use the NDA/BLA as the basis on which to grant approval for the drug to be brought to market. Regarding the NDA specifically, the FDA states:

The goals of the NDA are to provide enough information to permit FDA reviewer to reach the following key decisions:

- *Whether the drug is safe and effective in its proposed use(s), and whether the benefits of the drug outweigh the risks.*
- *Whether the drug's proposed labelling (package insert) is appropriate, and what it should contain.*
- *Whether the methods used in manufacturing the drug and the controls used to maintain the drug's quality are adequate to preserve the drug's identity, strength, quality, and purity.*

Source: https://www.fda.gov/drugs/types-applications/new-drug-application-nda

Launch

Once market approval has been granted by the regulatory authority, the drug is ready for launch. By this point, the drug discovery and development cycle may have been going on for over a decade and could have cost in excess of $1 billion, so it is vital that the launch process is planned and executed with great care. In some cases, the planning will have already commenced prior to and during the submission process.

There are various considerations that need to be addressed leading up to a product launch:

- Assign appropriate pricing
- Choose a brand name
- Design packaging
- Produce a packaging insert – this will inform health professionals and patients on

the safe use of the drug and any potential side effects

- Manufacture drug on a commercial scale
- Ensure robust supply chain is in place so that the drug can be distributed effectively
- Engage with key stakeholders and raise awareness of the drug, highlighting the key benefits (especially over existing treatments) to patients. Figure 3 shows examples of key stakeholders to consider in preparation for a product launch

Figure 3: Some of the key stakeholders to consider leading up to product launch

- Digital marketing campaign – utilising online platforms (social media, online advertising, email marketing, etc.) to reach potential consumers and enhance brand reputation. The healthcare industry is subject to strict regulations, so appropriate care and attention must be paid when designing a digital marketing campaign

The process of taking a drug from the clinic to the market abounds with challenges, and so pharmaceutical/biotech companies often have entire teams or departments dedicated to the commercialisation function.

Chapter 6:
Post-launch surveillance

After successfully launching a new drug, there are further studies (classed as phase IV) that are sometimes conducted to monitor the safety profile of the drug once it is on the market. This is known as pharmacovigilance, and it provides valuable insight into the drug as it is used outside of a strictly controlled clinical trial setting, in a wider population over a longer period of time. Sometimes adverse side effects, or 'adverse events' as they are formally called, only become apparent post launch and are carefully documented and monitored throughout this phase. The rate of occurrence of a side effect may be too small to be detected in a clinical phase population size and may only become apparent when the drug is introduced to the market, making it more widely available to a larger population. The drug company must put measures in place for patients and physicians to report adverse events. The adverse events are reported to the regulatory authority and if these pharmacovigilance studies uncover new concerns about the safety of the drug then it may even be withdrawn from the market pending further investigation/trials.

Another consideration is drug-drug interactions. Sometimes a drug can interact with another drug that the patient is concurrently taking which can have a range of outcomes. One drug may bind to another and in doing so affect the proportion of either being available for its intended therapeutic action. Drug-drug interactions can also impact their metabolism and either diminish or enhance their actions. In some cases, this can cause unwanted side effects and again, such findings may only become apparent after the drug is brought to the market.

The requirement to conduct post-launch phase IV studies is often mandated by regulatory authorities, but sometimes drug companies perform these studies voluntarily. Aside from safety monitoring, there are other benefits that can be taken from phase IV studies; for example, the drug can be tested for other disease indications to further expand its potential use, thereby enhancing its commercial value. In addition, the aforementioned drug-drug interactions can be studied specifically to demonstrate the benefit of taking a combination of drugs. This is common for cancer related therapies where drugs are given in combination to enhance the killing of tumour cells. It should be noted that such drug-drug interactions are sometimes tested in the pre-launch clinical phases as part of the study designs.

Chapter 7:
Generics, Biosimilars and Biosuperiors

Pharmaceutical drugs are usually patent protected for a period of 20 years, during which the drug cannot be copied; however, the patent duration does not begin at the point of product launch but rather earlier in the drug discovery and development process. This means that by the time the drug reaches the market, a substantial portion of the patent period has already passed, and it may only have market exclusivity for a few years thereafter. Once the patent expires, other companies can produce and market copies of the drug under a different brand name, which almost immediately reduces the value of the original drug. The copies of the drug are called generics, which are chemically identical to the original drug and can appear on the market very rapidly once the patent for the original drug expires.

On the other hand, large molecule biologic drugs are more difficult to copy as they are more complex in nature and produced in biological systems, which means it will be near impossible to create an exact copy. For this reason, molecules that are created to imitate large molecule biologic drugs are called biosimilars. Biosimilars, therefore, are required to be subjected to clinical testing in order to demonstrate that they are indeed equivalent to the original biologic drug.

Sometimes an improved version of a biologic drug is developed so that it has certain advantages over the original. Such biologic drugs are called biosuperiors (or biobetters) and they may, for example, have a slightly different structure or a change in the formulation that makes it more effective than the original biologic drug. This could mean that the treatment is administered at lower doses or at less frequent intervals, thereby improving on the cost and convenience. As with biosimilars, biosuperiors are subject to clinical testing before they are approved for the market.

PART 2

Chapter 8:
CMC

One of the key elements of the drug discovery and development process is CMC, which stands for chemistry, manufacturing and controls. The purpose of CMC is to establish the procedure for manufacturing the drug, to assess its characteristics/physiochemical properties and to carry out appropriate stability testing in order to ensure that the drug is produced to the highest safety and quality standards. This also helps to deliver consistency between different batches of the drug that will be used throughout the drug development process. CMC is an ongoing process that strives to not only maintain the consistency of drug molecule production but also to improve it where necessary. Three key areas of CMC are characterisation, stability and formulation:

Characterisation – This helps provide an understanding of the physical and chemical attributes of the drug, which ultimately affect its appearance, stability, ability to be processed and its overall performance. Some characteristics of the drug that are determined/analysed include its size, affinity for its target, osmolality and purity. The methods involved in characterisation undergo formal development and validation to ensure the robustness and integrity of the results generated.

Stability – Rigorous testing of the stability of the drug is performed under various conditions in order to establish expiry/retest dates to ensure that the integrity of the material is intact when it is used. Determining the stability of the drug at different temperatures is particularly important as it will ultimately help to define the storage condition of the final drug product. Commonly the stability is tested at frozen, refrigerated and room temperatures.

Conditions relating to the shipment of the drug must also be tested, and this includes exposure to heat, light and the length of time that it may be in transit when it is shipped between testing sites.

Formulation – A formulation consists of different chemical substances combined with the active drug, which result in a medicinal product. A suitable drug formulation must be designed in order to ensure that it is stable and can be administered to the patient safely and effectively. Drug formulation is a vital discipline within the drug development process and is refined throughout the process as scientists gain more of an understanding of the drug, sometimes even beyond launch.

Consideration must also be given to the pharmacy handling procedures, where relevant. For example, if the drug is to be administered to the patient via saline bags, then every aspect of this procedure, from the specific bags to be used and the exact procedure for filling the saline bags with the drug to the procedure for administering the drug to the patient, must be assessed and standardised so that there is minimal risk of variation in the way the drug is administered to patients.

Drug companies tend to have an entire department dedicated to CMC, but in some cases the CMC activities are partly outsourced to CRO partners. When the drug is ready for large scale manufacturing for use in clinical trials and for commercial purposes post launch, the drug company can collaborate with a CMO (contract manufacturing organisations), which specialises in the manufacture of drugs to the highest regulatory standards. Building the infrastructure to be able to manufacture drugs at such a large scale and to such strict regulatory standards can be extremely time consuming and costly, which is why CMOs are commonly used as an alternative.

Chapter 9:
GLP/GCP/GMP regulations

Any industry that is involved in producing goods for human consumption needs a level of regulation to ensure that the health and safety of the consumer are put first. So it is with the pharmaceutical industry, which is one of the most heavily regulated industries in the world. In the drug development cycle, the regulatory standards are applied at the preclinical and clinical testing stages and during the manufacture of the drug when it is ready for use in human trials. Each stage is governed by a different set of regulations, although there is some overlap of the guiding principles of each set.

GLP

GLP (good laboratory practice) is a set of rules that is applied to the conduct of preclinical studies and is designed to ensure the accuracy and integrity of the data generated in these studies. This is necessary because data generated in such studies will aid the design of early-phase clinical trial studies in which the drug will be administered to humans for the first time; therefore, it is vital the data is accurate and reliable. GLP regulations provide a framework to ensure that preclinical studies are planned, performed, monitored, recorded, reported and archived. Any site that is involved in the conduct of a GLP study must have GLP accreditation, which is granted by the regulatory authorities following a formal inspection to see if the site has all the infrastructure in place to conduct GLP studies. Some of the key aspects of GLP compliant studies are:

➤ The study is under the control of a study director, an individual who is responsible for the overall conduct and compliance of the study. For multi-phase studies – for example, when the in-life phase is conducted at one site and the laboratory testing of the samples is shipped and performed at another site – principal investigators can be assigned to take responsibility for different phases of the study. However, the study director will still be responsible for the overall conduct of the study.

➢ The study is conducted in accordance with a study plan (approved by the study director) that outlines all aspects of the conduct of the study and how it is to be recorded and reported. Any unplanned deviation from the study plan is recorded and the impact of the deviation is assessed. On completion of laboratory activities, the study director will then compile a study report. For multi-phase studies, the principal investigators will write their phase plans and reports pertaining to the particular phase of the study that they are responsible for and these will be incorporated into the final study report.

- ➢ Various aspects of the study are subject to a QA (quality assurance) audit. QA is a function that monitors compliance of the study to GLP regulations and will audit the study plan, the laboratory testing of study samples and the study report. The audit involves detailed scrutiny of each aspect, ensuring that the study is planned, conducted and reported in accordance with the GLP regulations. QA will also inspect test facilities (where the study director is located) and test sites (where phases of the study are conducted) to ensure they meet the necessary GLP standards. A drug company may have their own in-house QA department; alternatively, this function can be outsourced to a QA consultant.

➢ On conclusion and issue of the report, the study is appropriately archived so that the information generated from it can be safely retrieved at any time. The regulatory authority may request that a study be retrieved from the archives as part of a routine inspection that GLP accredited sites must undergo.

If a preclinical study does not meet GLP regulations, the regulatory authority may consider refusing to accept the study as part of a regulatory submission.

Although GLP did not originate in the US, it was adopted by the country following an ill-famed scandal in the 1970s in which one the country's most prominent industrial product safety testing laboratories at the time (Industrial Biotest) was found to have been conducting fraudulent studies on the safety testing of household and industrial products. This case led to the US adopting the GLP regulations in order to improve the standards of chemical and pharmaceutical testing and minimise the potential for fraudulent conduct in these studies there onwards.

GCP

GCP (good clinical practice) is an international set of ethical and scientific quality standards by which clinical studies are planned, conducted, recorded and reported with particular emphasis on the rights, health and safety of study participants. Some of the key principles of GCP include:

> ➤ There should be clear and detailed scientific rationale for any research involving humans and there should be preclinical data to support this.

- ➢ The overall potential risk/benefit to the study participants should be outlined beforehand.
- ➢ Prior independent ethics committee (IEC) / institutional review board (IRB) approval should be obtained.
- ➢ An approved clinical study protocol should be in place, and all work should be performed in accordance with it.
- ➢ Any study participant must freely give their informed consent.
- ➢ Individuals involved at any stage of a clinical trial must be appropriately trained and qualified in their role.
- ➢ The privacy and confidentiality of the study participants must be strictly protected.
- ➢ All information generated in the clinical trial must be recorded promptly and accurately and should be subject to rigorous review.

> ➤ Any investigational drug product that is used in the clinical trial must be manufactured, stored and handled in accordance with GMP (good manufacturing practices).

> ➤ There should be appropriate QA oversight of clinical trial studies.

GCP compliance is mandatory not only for pharmaceutical companies but any organisation that may be involved in the clinical trial studies such as CROs, hospitals, universities and so on.

GMP

GMP (good manufacturing practice) in drug development is a set of standards to which any manufacturer of a medicinal drug product must adhere in order to ensure that the product is of the highest quality and that there is consistency when different batches of the product are manufactured. It also ensures that the product is suitable for its intended use and that it fulfils the requirements for clinical trial or marketing authorisation. Some of the key principles of GMP include:

➤ The manufacturing facility must have high cleanliness and hygiene standards and must maintain controlled environmental conditions to ensure the protection of the manufactured product from contamination.

➤ The manufacturing must be well defined and controlled. To ensure consistency between batches, the manufacturing process needs to be validated.

➤ All procedures and instructions must be written clearly and in adherence to good documentation practices so that they are easy to understand and follow.

➤ Any information recorded during the manufacturing process must be done so promptly and accurately, and all records should be reviewed. Any deviation from the process must be recorded and the impact assessed.

➤ Any individual involved in the manufacturing process must be appropriately trained and qualified.

➤ There must be appropriate facilities in place to recall a batch of product from sale or supply if necessary.

All of these principles help to ensure that the end product meets the highest quality and safety standards before it is made available for use.

Chapter 10:
Small molecule drugs vs large molecule biologics

There are two types of drugs that the pharmaceutical industry primarily focuses on: small molecule drugs and large molecule biologic drugs. Most small molecule drugs are chemically synthesised, while large molecule biologics are generated in living systems such as cell cultures and are more complex by nature. This also means that any small changes in the manufacturing process can have a larger impact on biologics, which could alter its efficacy and safety properties. Small molecule drugs are more numerous as they have been the focus of the pharmaceutical industry longer, but biologics are becoming more and more prevalent.

Figure 4 outlines some of the differences between the two types of drugs.

	Small molecule drugs	Large molecule biologics
Size/molecular weight	Small, typically <1 kilodalton	Large, typically up to hundreds of kilodaltons
Structural complexity	Simple	Complex
Production	Chemically synthesised	Produced in living systems
Typical site of action	Intracellular	Extracellular
Typical route of administration	Oral	Intravenous/sub-cutaneous injection

Figure 4: Some of the differences between small molecule drugs and large molecule biologics

As each type of drug has different properties and characteristics, there are variations to their respective journeys through the drug discovery and development process, even though the overall path to the market is similar. The differences in the generation of each type of drug (chemical synthesis for small molecule drugs and expression in living systems for biologics) mean that there is a lot of contrast in the early discovery phase between the two. In addition, during preclinical and clinical development phases, the approach to bioanalysis is different for each. Small molecule drugs require the use of technologies such as LC-MS/MS (liquid chromatography with tandem mass spectrometry) and HPLC (high-performance liquid chromatography), which are more suited to analysing smaller molecular weight compounds. On the other hand, bioanalysis of large molecule biologics is more commonly performed using ligand-binding assays such as ELISA (enzyme-linked immunosorbent assay).

Chapter 11:
Immunogenicity

One important consideration for bioanalytical testing during the clinical trials is immunogenicity. Immunogenicity in its simplest term is the ability of a foreign substance to provoke an immune response when introduced to the body. Therapeutic drugs are foreign substances that can cause such immune responses, although depending on the nature of the drug, the probability/extent of this can vary. For example, a therapeutic that is a recombinant version of a naturally occurring protein may be (although not always) less likely to provoke an immune response than a therapeutic that is chemically synthesised and not naturally occurring in the body. One of the ways immunogenicity manifests itself is in the form of antibody production against the drug (known as anti-drug antibodies, or ADA), and these

92

antibodies have the potential to cause various side effects, some of which can be severe, for example, cytokine release syndrome or anaphylaxis. Therefore, it is a regulatory requirement to assess immunogenicity at the clinical development stage. ADA can also impact the PK/PD profiles of individuals by binding to the circulating drug, resulting in a change in the rate at which the drug is normally cleared from the body or by binding to an active site on the drug and thereby preventing it from binding its target, thus negatively impacting efficacy (such antibodies are known as neutralising antibodies).

The immunogenic potential of a therapeutic, that is, how likely it is to cause an immune response, can be influenced by various intrinsic and extrinsic factors as outlined in figure 5.

Extrinsic	Intrinsic
❖ Formulation	❑ Degradability
❖ Route of Delivery	❑ Structural Complexity
❖ Frequency of Administration	❑ Similarity to self-antigens
❖ Dose Size	❑ Immune status of the Patient
❖ Aggregation	
❖ Patient Population	
❖ Other Medication	

Figure 5: Factors that can influence the immunogenic potential of a drug

The occurrence of ADA response to therapeutic drugs is not uncommon and it is not always a serious cause for concern because sometimes they have no discernible impact. Even if the ADA impact the PK and PD profile of the individual, this can in some cases be addressed, for example, by adjusting the dosing concentration or frequency to overcome the ADA effects.

Chapter 12: Pharmaceutical/biotechnology vs CRO – a personal perspective

In this concluding chapter, I want to compare and contrast what it is like to work at a pharmaceutical/biotechnology company and a CRO (contract research organisation). Having worked at both types of companies over the past 13 years in a bioanalytical function, I feel that it may be valuable for readers to know some of the differences that can be expected in working at one versus the other, since both provide opportunities for individuals with the necessary skills and qualifications to work in this industry. I should add that this is my personal experience in one particular function, so it may not necessarily reflect the experience of others who have had similar such opportunities.

When working at a pharmaceutical or biotechnology company, the objective is to play a role in advancing the company's drug molecules through the drug discovery and development pipeline. There is an abundance of activity that occurs at each stage of the drug discovery and development process, and it is likely that you will specialise in a role at a particular point in this process. The ultimate goal of the company will be to bring drugs to the market and in doing so generate profit while bringing benefit to patients. As shown in part 1 of this book, the journey from discovery to market can take a number of years, so it can require a lot of patience to be able to experience the success of reaching this long-term goal. Of course, you may join a company at a time when it has a rich pipeline of drug candidates at every stage of the development cycle, so the wait for success may not always be long. I have also found that there are plenty of opportunities to work with colleagues in functions upstream and downstream of my own, which has allowed me to continuously learn about the entire drug development process even though my focus is in one particular area. Some pharmaceutical and biotech companies focus on one particular disease area such as oncology or respiratory, but in my time at MedImmune and at Immunocore I have had the opportunity to work on drug projects from multiple diseases areas and it has been a valuable learning experience.

Working at a CRO means that you are providing a service to a pharmaceutical or biotechnology company. There is an emphasis on the day-to-day generation of revenue from performing contracted work, whether it be laboratory analysis or data/report generation, among other things. This is understandable given the business model of a typical CRO, but this does not distract from being involved in the science. I have found that working at a CRO allows you to interact with multiple companies and to work on various different types of potential treatments. This meant that there was always variety in the work that I did, and it also gave me the opportunity to establish collaborative relationships with many scientists across the industry, which has been rewarding. I have found that there are different pressures when working at a CRO as I was always working with multiple clients, which made it difficult to prioritise the work for one client over another when needed.

In summary, there are benefits and minor downsides to working in either sector, but for myself it has been an enriching experience in both cases. The partnership between pharma/biotechs and CROs is a highly collaborative one so there is a lot of opportunity to learn about the industry in both situations.

Summary

The drug discovery and development process is long, costly and abounds with challenges at every stage, but the success of bringing a drug to the market can be incredibly rewarding, especially knowing that it will have a positive impact on patients' lives. Navigating a potential drug therapeutic through each stage of the process requires expertise in many scientific disciplines and a robust knowledge of the regulatory requirements, not to mention the commercial know-how for a successful product launch. There are various regulations in place to ensure accuracy and integrity of data generated in preclinical and clinical studies, and to safeguard the rights, health and well-being of all patients and volunteers. This means that there are many opportunities for careers in the industry in a broad variety of disciplines. I hope that this book has been valuable in providing a relatively brief overview of a very long and complex process.

Thank you for reading this book, I hope it has been beneficial to you. I would greatly appreciate if you could take a few minutes to provide feedback by reviewing this book.

About me

I graduated in 2007 from the University of Bristol with a BSc. in Cancer Biology and Immunology. Since then I have worked at several companies in a bioanalytical role spanning 13 years, as of writing. I began my career at Covance Inc. before moving on to MedImmune (now retired in favour of AstraZeneca, of which it was a wholly owned subsidiary), Eurofins Pharma Bioanalytical Services and Immunocore, where I currently work as a senior scientist in the Bioanalytical team. You can find more detailed information about my previous and current roles on my LinkedIn profile:

linkedin.com/in/kabir-hussain-b1371951

Printed in Great Britain
by Amazon

61143186R00061